The Genial Seed

The Genial Seed

A Japanese Song Cycle

Translated by Frank Hoff

A Mushinsha Book
Grossman Publishers

The Genial Seed is a translation of a mediaeval Japanese cycle of songs intended to accompany the annual rice-transplanting festival. The manuscript translated, called *Tauezōshi* in Japanese, is in the Japanese Language Study, Department of Literature, at the University of Tokyo. A page of an essentially similar manuscript in the Historiographical Institute of the same university is illustrated on p 173. The photographs of related contemporary festivals in various communities in the Chūgoku area of Japan were taken by Hagiwara Hidesaburō. The illustrations on pp 16–17, 110–111 are from a Momoyama Period scroll (artist unknown), called *Tawara kasane kōsaku emaki*, in the possession of the Historiographical Institute. The illustration on p 60 is from a scroll owned by the Aso Shrine in Kumamoto Prefecture. Another scroll, called *Rōnō yawa* and dated 1843, is reproduced on the jacket. The illustrations on pp 94–95, 146–147 are by Ishikawa Ryūsen, and are taken from a Genroku Period book called *Yamato kōsaku eshō*. The originals of the bull course diagrams (*shirokaki*) reproduced on the end-papers are in the collection of Yamamoto Satoichi in Hiroshima Prefecture.

First published in the United States of America
in 1971
by Grossman Publishers, Inc.
44 West 56th Street
New York, N.Y. 10019

Designed and produced by Mushinsha Limited
IRM/Rosei Bldg., 4, Higashi Azabu 1-chome, Minato-ku
Tokyo, Japan

Printed in Japan

First Edition, 1971
Library of Congress Catalog Card No. 76-145996

Contents

animos tollent sata

VIRGIL

The womb of earth the genial seed receives:
For then almighty Jove descends. . . .

VIRGIL (JOHN DRYDEN)

So closely, indeed, is the practice of this particular art interwoven with . . . everyday life, that singers, unable to recall a song I had asked for would often make a remark such as: 'Oh, if only I were driving the cows I could sing it at once.'

CECIL SHARPE

'If Christ does not rise tomorrow we shall have no corn this year'

Reported by JANE HARRISON

. . . they deal after their own fashion, to the honor of Baalat and with a feast of pairing, with cymbals and shouting, for the sake of the fruitfulness.

THOMAS MANN

alternis dicetis. amant alterna Camenae

VIRGIL

Preface

THE TRANSLATION in the following pages is of a sequence of song once sung in the fields during the transplanting of rice-seedlings at the end of May or in early June in certain areas of rural Japan. The sequence is a form of song developed in part to meet the requirements of this agricultural event but is perhaps best read today as a single, long poem. The conditions of its performance gave it a unity but there is also great variety in its contents. Produced within a single ritual context, it drew its material from a number of different traditions of song, poetry and even storytelling. It is thus related, on the one hand, to other texts for performance in a ritual setting, literary texts which took their unique shape within a performance, usually a festival, context. On the other hand, it is related to songs of a variety of types from the same and earlier historic periods. I would like briefly to look at these two aspects of this sequence of songs.

In the district where it originated, Ōasa Machi, Shinjō in Hiroshima Prefecture, something like the event which produced this text is still performed, though in greatly abbreviated form, and now largely as a spectacle. However, at other places in the five prefectures known together as the Chūgoku area large-scale ritual transplanting events, *ōtaue* or *hanadaue*, are still a genuine and important part of the social life of the people. There follows a brief composite view of this type of transplanting event in which I have greatly simplified the degree of local variation.

Well before dawn workers gather in a special shelter. At one time there was general talk here of a traditional nature including talk about sex that had taken place during the night just ending. In this way sex begins to play its important role in the day's ritual even before the songs referring to it are sung later in the morning. At some places seedlings are gathered early in the morning from the nursery-field where they have grown from seeds scattered some forty days before. (Seedlings are gathered to the accompaniment of a special group of songs, *naitori uta*, not included in this translation.) In some areas seedlings are gathered not once but several times throughout the day to replenish the supply as the transplanting girls use it up. Before the actual transplanting begins a ritual invocation of the field god is performed. The participants march out to song and music and gather before a temporary shelf set up in a field or on one of the narrow lanes between fields. The invocation is called *sanbai oroshi*. An early form of this rite of calling down, or manifesting, the field god, *sanbai*, may simply have been to insert a tub into the center of a field or to place three rice seedlings there or on a stand. At present, the *gohei*, that familiar sacred white Shinto streamer, is a part of the religious decorations on most *sanbaidana*, as this shelf is called, though Buddhist decorations also have a place. The genealogy and an account of the birth of the field god is sometimes read as a part of the ceremony. Otherwise, an ordinary Shinto prayer is recited.

While the transplanting girls gather for this ceremony of invocation in a field where a ritual plowing has just been completed, the plowing and readying of a second field may begin. Though local traditions and the topography of the area in some degree determine the actual order of events, a ritual plowing is a part of most transplanting events of this special type. Prize bulls are decorated and led into the fields; diagrams may even prescribe the patterns which the line of bulls should make as it passes over the field: an ordered maneuver is necessary to keep animals out of each other's way in the narrow fields, but the patterns themselves may be significant. In any case, configurations for the ritual often differ from those used for normal plowing. Many have names: '*sanbai's* rope,' the pattern followed at the morning's invocation, *oroshi*; 'a sleeve of one-ply;' 'a sleeve of two-ply;' 'the dancing crane;' 'crane in his nest.' Some are prescribed for a certain time of day or for a field of a certain shape. The shape of a field may even determine if it is suitable at all for the field god's epiphany.

Bulls are prized for their strength and beauty. It is an honor for a man to be chosen to lead one of the key animals. The orderliness of the complex maneuvers in the field is in the hands of these drivers. One reason why this type of event is becoming scarce at present is that with mechanization there is less need for, and, consequently, less interest in keeping prize bulls. The decorations on the bulls are costly. The line of animals, parading into the field and then proceeding with the plowing, is a gorgeous sight. Each animal carries the name of the house he represents. Keeping an animal for the festival is an expense that only the wealthy farmer can afford. A costly public display of this sort is an example of how the economy of the community and its festival are interlocked.

The workers too parade into the field from the community shelter where they have gathered before the *sanbai oroshi*. They march to the accompaniment of special music and song. The day's events are a splendid show as well as hard work. The two aspects must have been mixed together from the very beginning, though, historically, increasing affluence allowed a larger scope for display. A large-scale and costly transplanting event of this sort is traditionally reserved for special fields: that of a large landowner or one closely associated with the god, the god's field, *shinden*, and because of the sacred nature of the event young women or girls do the actual work of transplanting.

After the field has been plowed by a line of bulls, men with rakes go in to break up clods and smooth it over before the transplanting. Once the transplanting has begun seedlings gathered from the nursery field are delivered to the planting girls as they need them.

The seedlings are inserted into the mud by a line of transplanting women, *saotome*, which moves backwards across the field. The leader, *sanbaisan* or *sagè* as he is called in one area, stands before the row of *saotome*. He beats two bamboo sticks, *sasara*, together; in some areas he rubs one stick over the zig-zag surface of another. Drums, flute, and small cymbals played by young men in the field provide the music to which the *saotome* and the leader sing the songs. The leader sings the first line, the parent song, *oyauta*; the *saotome* answer with the second, the child song, *kouta*. Together the *saotome* and the leader sing the following lines, *oroshi*, of one stanza and then go on to the next.

The field work is hard; there are frequent rests. There is a main break for food in the late morning. Food is still brought into the fields for the workers. Several genre paintings from the late middle ages or early modern period show food and drink being brought out for a large-scale transplanting event. At one festival in Kyushu today women dressed in white carrying boxes like those shown on the genre screens still preserve this ancient custom (see stanza 28). Tradition relates these women to the god of the field: in a sense both are mysterious visitors from afar bringing bounty, who return and so have to be seen off, as the songs say, after they have bestowed their gifts.

As the work draws to a close and it becomes dark the workers' thoughts turn again to the enjoyment of sex which has traditionally been a part of the event. The poetry of their songs reflects this. In some places the *saotome* scatter mud over the young men. The men have not been doing the transplanting. Their work is preparing the fields, delivering the seedlings and providing the musical accompaniment for the singing.

In brief this is the type of ritual event within which the long set of songs translated here was developed.

IN SUCH a ritual event, obviously the conditions of performance, both ritual and work, shape the songs into stanzas, sets, and the larger whole. The same factors determine what we might call the meaning of the poetry. We see this clearly if we compare songs outside the ritual setting to the use made of the same or similar songs within it. In making the comparison we also realize how many individual lines and phrases are shared by the transplanting tradition with types of poetry and song independent of it. The largest body of poetry sharing elements with the sequence translated here is known by Japanese scholars as *kayō*, which we might think of as performed poetry or song. And, although the sequence is certainly the product of its ritual environment, another way of looking at its poetry is as a gathering together of various strains of *kayō* from the Japanese late middle ages (1333–1600) and the early modern period (1600–1868).

Stanza 19 of our sequence refers to the planting girls' long hair, and the immediate group (stanzas 15–19) develops the theme of hair and other aspects of personal appearance. Hair is repeated in the fourth set for the evening (stanza 120) where it is definitely a woman's hair tangled from sleep with a man, *nemidaregami*. The fact that a motif is repeated in the sequence is itself significant. (The sliding door of stanza 1 reappears in stanza 122 for the evening where the lover returns.) It seems as though an expression like long hair, dark hair, trailing hair, tangled hair, finely combed hair was almost required among morning songs within this special tradition of transplanting songs. The expression *nemidaregami* was in use at an earlier period in Japanese poetry and later in the early modern period it became especially popular. However, several facts, including the consistency of its use in related transplanting sequences, suggest that the motif of hair is used in a special way here. The frequent mention of hair is a way of encouraging the rice crop itself to bear long, rich, lustrous tassels, and so yield a full crop. What the Japanese call *juka*, poetry as magic, seems to be one major 'meaning' of this long sequence.

The last line of stanza 120:

> Never ever to forget that look, her hair still matted from our sleep together

appears in a number of separate song types popular in the later middle ages and the early modern period. Although there has been research tracing how a passage probably moved from one sub-type to another, what I want to emphasize is that the popularity of a line like this, and it is only one example to which more could be added, shows how closely associated a number of popular genres were at the period. These were the sources of much of the poetry which has found its way into this long sequence for performance in the field. In order to show how widespread was the network of *kayō* at the period, I would like to indicate here a number of texts sharing the concept lying behind this single line; with the exception of the first, *kusemai*, all are examples of the particular type of *kayō* which is most representative of the period, what is called the short song, *kouta*.

Kusemai, popular in the early Muromachi period (1333–1600), was a performance of some kind of dance or movement to accompany song. This form was adapted for Nō by Kan-ami as one major step in its development some time before the last quarter of the 14th century. A *kusemai* number called *Jishu*, or *Mizukumi*, contains five *kouta*. These same five songs were later collected into the important anthology of *kouta* called *Kanginshū* (1518.) Among the *kouta* worked into the *kusemai* number is this popular song:

> The sash to my dress, brocade like flowers
> I opened it, but what did I get
> My heart is confused like the branches of the willow
> When will I ever forget how you looked your hair still tangled

Versions of the *kyōgen* play *Hanago* contain a *kouta* with the same elements as the song included in this *kusemai* number and as stanza 120 of the transplanting sequence. The hero of the play sings several songs describing the bliss of his love-making with young Hanago. The song in question is included. There was a close relation between the *kouta* tradition and *kyōgen*. In fact, many plays seem to supply just enough plot material to create a convenient stage context for singing *kouta* accompanied by dance.

A collection of popular songs made sometime between 1592 and 1615, called the *Ryūtatsu koutashū*, also contains a song related to the line from stanza 120. Sets of song for the early Kabuki, again songs for dance, are collected in a work which dates to 1596–1630. Among these one song sequence, *Yayako*, contains a similar song. A later collection (1703) of song sets for samisen and dance, *Matsu no ha*, has a song, *Hanagasa*, with a line about black hair tangled in sleep.

This should illustrate some of the kinds of poetry outside the sequence with which transplanting songs were associated and may serve, in a sense, to define what is characteristic of *kayō* in the late middle ages and early modern period when the sequence was being formed: it was the period of the short song, or *kouta*. Earlier in the middle ages, during the Kamakura period (1185–1333) a longer form, *sōga* or *enkyoku*, recited in performance within the military class, was the major genre of *kayō*. Still earlier, from the end of the Heian into the early middle ages, the *imayō* song was its mainstay. Songs of this type make up a large part of the collection called the *Ryōjin Hishō*, and several are translated by Arthur Waley, as well as in the *Penguin Book of Japanese Verse*. Donald Philippi has translated into English most of the *kayō* from the early period in his *This Wine of Peace, This Wine of Laughter*. Noah Brannen in his '*Kinkafu* Collection of Ancient Japanese Song,' *Monumenta Nipponica* vol. XXIII, No. 3–4, has translated and given an introduction to *kayō* collected in the early Heian period.

The reader will himself realize how varied is the content of this sequence of transplanting songs; this brief mention of some related *kayō* suggests in part how this variety was possible. Many songs in the sequence came to it from several smaller groups within the history of *kayō*. Both in concept and expression a few are close to *kayō* from the period before the middle ages. Agricultural songs from the later Heian as well as sacred songs, *kami uta*, from the same period and from the middle ages have all, in one form or another, entered into this long sequence, especially noticeable being the traces of *kayō* from the later part of the middle ages, the so-called *kouta*.

But it would be a mistake to think of the song sequence as a mere random compilation, a picking up of a song from here and another from there. Rather it is a single whole: *kayō* meant to be sung on one single occasion. *Ōtaue* is an occasion for which songs from outside sources have taken their own place, an easy and harmonious one, beside songs within the more strictly speaking direct tradition, including extemporaneously composed ones, of the transplanting ritual. Contributing elements have taken a place within a new whole, acquiring meaning within the total context of a single ritual structure, *ōtaue*. We can imagine the process of the formation of such a sequence as one in which songs from a number of *kouta* types, as well as various sacred songs, have been drawn over a period of time into a new and larger song whole, that of *kayō* performed on the occasion of transplanting. And within the new context they have acquired new life.

THE AMALGAMATING power, what has given cohesion here to short songs coming from various sources, is, in part, the ritual event, *hanadaue*, itself. We get a larger perspective on it as an environment supporting the growth of a long sequence of song by extending our view to include other festival types in Japan. We can do that succinctly through the classification of folk performance types made by Professor Honda Yasuji of Waseda University. Professor Honda's concern in formulating this classification was not the question of literature in a performance setting: his orientation is not literary; but neither is it anthropological—it belongs to the specialized Japanese discipline of the study of folk performance, *minzoku geinō*, and his tripartite division was a way of organizing the extremely large body of actual performances he had observed himself. Though based upon performance in festivals, *matsuri*, which could have been seen in Japan during his lifetime, his classification has made use of historical knowledge as well: written records, for example, about folk events, including those which developed into what we now consider the classical forms of Japanese theater. There is no need to assert that every event within his classification has produced a recorded text of literary value; nevertheless it is a useful scheme for locating the work translated here in one of its important dimensions.

Kagura, as defined by Professor Honda, is an event performed for the prolongation of life. Events known locally as *kagura*, or placed by Honda within this type, are found today both in the imperial court and in the countryside. Events of this type go far back historically, at least to the Heian period (794–1185). For instance, it was in the Heian period that a series of songs for *kagura*, *kagura uta*, took the shape we are familiar with now. If we view these songs as a single sequence, *kagura uta* is an early example of an instinct within ritual toward an integrated series composed of otherwise unrelated songs. One reason for performing this type of court *kagura* was (and still is, because court *kagura* is still performed today) to empower a rod or something held in the hand, *torimono*, during dance with divine power. Songs describing the physical or the divine aspect of a *torimono* accompany the dance. When grouped together these songs, in many cases 'folk' songs popular elsewhere outside the ritual context, compose an extended text produced for court *kagura*. It is interesting to note the similarity between one of these songs sung for dance with a *torimono*, in this case a bow, and one of the songs in this sequence. Song 82 (Bow that I love. Girls, if you pull—come, come) is related to this *kagura uta*:

In the province of Adachi
Of Michinoku
There are good trees fit for a bow
If I pull one it will draw
You, too, beloved, quietly now, quietly
Come to me forever

Another example of song associated with *kagura* in the larger sense used by Honda is a type known as divine songs, *kami uta*. At a later period *kami uta* developed into a song form of set stanza pattern. But more broadly applied the term refers simply to a song sung to manifest a spirit. The song translated on page 175 to invoke the god of the field is a type of *kami uta*.

Furyū, another of Honda's categories, was especially popular late in the Japanese middle ages and in the early modern period. *Furyū* is a way of driving off pestilence with costume, song and dance. A set of lyrics for dance was assembled from otherwise unrelated songs for performance in the context of a singly *furyū* event. The popularity of dance in the period accounts for the wide circulation of many of these songs or song sets. In fact, dance served as a way of transmitting popular songs from area to area. We often find a song referring to affairs in the center of culture preserved in widely separated areas and in quite remote country places. The mechanism of popular dance may explain how some of the songs in the long transplanting sequence translated here first came to be known in the areas where they were to become elements of such sequences.

The last of Honda's categories, *dengaku*, is the most relevant of his three types to this sequence. Honda believes that music, perhaps just the beating of a drum, was important in transplanting from an early time as a way of stimulating the growth of seedlings inserted into the field. Together with Professor Manabe Masahiro's view on the importance and early development of the use of a question-and-answer form in the fields this suggestion of Professor Honda's offers an explanation of the full development later of a ritual of song and music—drums, flute, *sasara* (bamboo sticks)—which became known as *ōtaue*, large-scale transplanting, the immediate ritual context within which the sequence translated here took shape. Honda's view is a generalized one: any event, not just music of the sort which accompanies some rice transplanting today, is appropriate for performance in the field to insure the effectiveness of planting. Generalized in this way his view suggests a reason even for the use of simple masked events, or dances, which were sometimes associated with transplanting. If these too were simply ways of stimulating growth, then some light is shed by Honda's view on the development of Nō plays, for in their formative period at least these were a part of the repertory of groups called *dengaku* performers. In any case, in *dengaku* performed in the field at the time of transplanting, ritual provides a context for the growth of a literary or dramatic art.

Each of Professor Honda's three types—*kagura, furyū, dengaku*— embraces a number of sub-types. For example, the present sequence belongs with a sub-type meant for performance at the time transplanting is actually done in late May or June. Another group, *taasobi*, is performed at New Year's, well before any field work is actually possible.

Taasobi is a type Honda calls 'prayer-in-advance.' In *taasobi* too a sequence-making instinct is active. Generally, the events of this ritual, and the words which accompany them, depict, or imitate, a full year's cycle of the farmer's work from the time before planting when he goes out to look over his fields through the harvest and garnering.

IN THIS Preface I have touched upon a number of points to explain how such a sequence of song could have come together in the first place. But though topics like the relation of transplanting to other rituals in Japan, or the relation of a *kouta* outside this sequence to its function within are in themselves interesting enough, they do little to explain either a contemporary's interest in the sequence as he found it in its original language or why he should have cared to translate it into his own. I translated these songs primarily to see if the result could be read in English as a single, long poem, which is the way I read them in Japanese. At first, perhaps, I saw the sequence too much as a novel: the sun slanting into a room where two people lie together and then at the end the same two people, I imagined, brought together once again in the dark, the sun recurring from time to time throughout the sequence as a theme of change.[1]

When it came to individual stanzas I noticed an effect almost of conversation—again perhaps I thought of a fragmented novel, perhaps a movie—two voices or a series of voices questioning and answering. The first line and its reply, what I later learned are called 'parent song' and 'child song,' bring up a subject, for example the sound of a bird calling as it darts across the field.[2] The remaining lines of the stanza can be almost random association on the theme spoken as if by answering voices. Or, quite a different effect, a stanza may suggest a dramatic encounter—a girl waits for a man's return from the Capital; neighbors defend a daughter-in-law against her mother-in-law's vituperation; a boy kicks a ball, a girl gives him tea to drink when he is tired.[3]

The grouping of stanzas into a set seemed to have its own unity. For example, the second set for the morning

[1] Stanzas 1, 68, 85, 93, 109, 117 and the group 126–132

[2] Stanza 24

[3] Stanzas 74, 100, 113

with its mention of haze, mist, preparing for work, remembrance of love, dew, parting in the morning, young inexperienced planting girls seemed to be loosely organized into some sort of whole relating to the time of day when it was sung. As I recall now, features such as these made me increasingly interested in this sequence of poems—for soon I was quite convinced that they were in fact poetry of a good degree of complexity. They are translated now and I expect that other readers will be interested for similar reasons.

When I knew something of the conditions of performance and the conventions within which the sequence took shape I found some of my earlier impressions sustained. For example, a simple question-and-answer form may, in fact, have been the prototype of the developed sequence. A tentative hypothesis to this effect has been proposed by Professor Manabe whose essay on certain questions raised by the manuscript tradition of the sequence comes at the end of this volume. He suggests that a question-and-answer form in the field at the time of transplanting—the social, religious and economic environment which supported the growth of the sequence—created some sort of magical *ambiance* within which rice seedlings would thrive. From that point of view, the developed sequence has merely elaborated the simpler question-and-answer form.

Again, something that had struck me at first reading, the larger pattern of the sequence—the succeeding events and impressions of a day from morning to night—came to be explained by the real nature of its composition. I no longer need to imagine the sequence as a fragmented novel. It is now fully meaningful as a sequence of song sung responsively by planting girls and their leader throughout a day's labor in the fields from early morning until evening, the contents of the poems varying according to the time of day, and including features of the landscape, animal life, ritual events, as well as aspects of the workers' everyday life.

The poetry of individual stanzas, too, acquires greater complexity with more knowledge of the conditions of this art. For instance, as I mentioned above, a simple poem like number 19 refers to the hair of the planting girls:

Sit along the veranda—the silver, the gold veranda
Wait a while to comb your hair
No, I won't comb it, I won't straighten it out
The man I love tangled it
I want to comb and set the sidelocks of your hair

The girls in the field bind back their long hair so that it won't interfere with their work. The series of poems from stanza 15 through 22 is about how the girls and the young men who play instruments look in the fields. How they looked during this social event was a point of interest when it was still an important part of communal life, because it was the occasion when the young and their parents looked around for marriage partners. In the poetry the beauty of country girls is idealized by references to standards set by refined tastes of Kyoto, the center of culture. (The manliness of the young men is idealized elsewhere by references to the warrior tales popular through recitation during the period.) In song number 19 a country girl who works in the mud all day is imagined resting for a while on a gold veranda. (This image suggests the eventual riches to come to the landowner himself at full harvest later in the year.) The girl's hair is matted. She doesn't care, however, because through poetry it is a lover's hands that have tangled it, not labor in the field. The mention of long hair is appropriate, as we have seen, in a ritual context where an abundant rice crop—forecast by long tassels hanging from the mature crop—is being, in a sense, prayed for. The presence of lover's hands is fully at home in poems still under the influence of the first set of morning songs with their strong message of renewal through sex.

The original text is a difficult one even for Japanese scholars. It would of course be doubly so for the English reader who wishes to understand it in all its detail. Nevertheless, because a full commentary would require many pages and would perhaps not so much increase the reader's enjoyment of whatever merit the translation may have as poetry, I have decided to limit explanation of difficult passages to those points which I felt were most important to the understanding of the sequence as a whole. For example, I do not identify the many place names which come up in it. An additional reason to withhold commentary at this stage is that what promises one day to become a nearly definitive commentary in Japanese, that of Professor Manabe mentioned above, is not yet fully available.

In my translation I have used a ". . . . ." to indicate words or passages whose interpretation is not generally agreed upon by Japanese scholars.

The Japanese text used for this translation, *Chūsei kinsei kayōshū*, (*Nihon koten bungaku taikei* #44), Iwanami Shoten, begins with the second set of morning songs. The English translation follows it in this regard. However, Professor Manabe in his essay argues that a set like that represented in the *Takamatsuya kohon*, translated on page 180, should be thought of as actually beginning the sequence as the first set of morning songs.

I want to acknowledge the assistance which I have received at various stages in studying these poems over several years from Yamaji Kōzō. Twice he read through the entire sequence with me, sharing his interpretations, both general and specific. In large part his view of this poetry has formed my own.

I am also greatly indebted to Professor Manabe Masahiro for his essay on the text beginning on page 169.

FRANK HOFF

Sophia University
Tokyo

The Genial Seed

Second Set of Morning Songs*

* For the missing First Set of Morning Songs, see Professor Manabe's essay, p. 180.

Sliding the rear door back a bit to see out* 1

Brighter than gold floods the morning sun

This morning's sun. How good its brightness feels, this morning's sun

Morning sun floods. It glitters upon the eastern hills

This morning around six looking out at the hills 2

Mist? Haze? Looking out at the hills

The mist is thick. It dances up around the flanks of Mount Misen

Clouds at dawn mean clearing later

Do they have rain in mind? Clouds around the flanks of Mount Misen

Put aside rain things—straw cape, hat. Set out. Clouds will clear

* The typographical arrangement of the stanzas is designed to suggest the first line (*oyauta*) sung by the *sanbaisan,* the second (*kouta*) by the *saotome,* and the remaining lines (*oroshi*) by both together; see Preface, p. 14.

3 Yesterday. Today too. What wind is blowing

If it is the love wind, gently

Bend. If you don't bend. Don't get tumbled by the wind

Don't spill it. Dew atop the broad-bell flower

Softly it blows. But the love wind pierces deep

4 Little rabbit from the farthest mountain, what are you after come all this way

Bush clover, cotton flower, dew. That is what you chase all this way

Making his way through bamboo grass. Dew. Not faithful?

He wants to. A faithfulness finer than thread

. . . . . small crow wet through with dew this morning 5

Ura-ura he cries on the wing wet through with dew

Bright and lovely is my master this morning

Closed in by morning mist young buck 6

No place to go he makes up poems

Pent in he can't make song nor poem

He took a tuft from the hairy belly of a deer and made a brush

He made a brush then used it up to write out the Lotus Sutra

7 Deep in a plowed field is where deer crouch closest to the earth

Deer in love cry lustily

A mere fawn cries like this: *konyo*

No, not a fawn at all. Crying with a threefold voice—king of the deer

First call of deer. Beyond, the sound of water over rapids

8 Don't sleep this morning. Get up, look out into the fields

See how the officer of the fifth month rowed deep into night

Fifth month's fifth month rain tangled hair after sleep, I suppose

Boatman just standing there on the rocky promontory

The end of her hair tangled round the hilt of his sword

Just hatched I suppose. Not yet even crawled out of your nest, warbler 9

Once you have all your feathers leave the old nest, warbler

I heard a voice but still inside the nest I guess—sleepy voice

Get some practice. Without practice it's a sleepy voice

The little one I've brought along with me hasn't yet got used to people

Looking over tea fields at Uji and at Toganoo 10

New leaves are out. They are at their best now

Shake off the dew. Pick them. New tea at the temple

I wish the girls had *kohon* pots. Let's pick tea

Let's pick the fresh tea, dew and all

II Pebbled beach in the distance and the salt flat there too

That's it. The fishermen's , the fishermen's
salt flat

They burn the sea weed and rake out the salt. They
let the sun dry off the beach

Level it out. Fine sand along the beach

Don't sleep. Look out at the white beach

Third Set of Morning Songs

The second month comes to Mount Yoshino 12

The mountaineer returns with flowers he has picked

Adorn yourself with buds from the cherry tree

They bring luck. Flowers are the seeds of things

Let's go pick flowers. To Mount Yoshino

Have flowers bloomed even among that band of planting girls from the hills 13

They have. They have. Bloomed with double blossoms

Lovely. Along the tips of branches flowers are in bud

The opening out of flowers is for the increase of his estate

The master of this field is called 'a man rich many times over'

14 The flower of indigo begins to wither. What shall we dye with it

Let us dye a dress for hunting, taking it to pieces, sewing it together again

Dyed and dryed. By dyers at Mount Shosha in Harima

Low class to dye a hunting dress a dark color

15 Buy me a Kyoto comb. A Kyoto comb's the thing

Make your hair graceful. A Kyoto comb's the thing

I've bought the comb. Now, comb your hair, young girl

The girl's hair is longer than she is

Long hair. If you walk it's liable to get caught in your sandals

Comb out your long hair then tie it back. If you don't tie it back 16

It smacks of what they do in the Capital, if you don't tie it back

Won't they say of unbound hair: 'hair tangled in sleep'

She combed out her long hair. It was beginning to be tangled in sleep

She tied it back. Her black hair longer than she

Serving at court keeps you busy. Next to the paper partition 17

Combing hair, powdering, rouging. Next to the paper partition

You like serving at court is what people will think, won't they

I wish I had the clothes to serve at court

Loan it to me. Your short-sleeve coat

18 The one beating the drum is Kudotarō. The one playing the cymbals is Kajiwara

Our own local Tarōs and Jirōs will have to fix themselves up too

I intended to get ready but forgot the things to do it with

Give me some. I want to slick back my hair. Give me some grease from the jar in your hand

19 Sit along the veranda—the silver, the gold veranda

Wait a while to comb your hair

No, I won't comb it. I won't straighten it out. The man I love tangled it

I want to comb and set the sidelocks of your hair

Fourth Set of Morning Songs

They dyed clothes for the planting girls to wear 20

They dyed them with the clover flower

Dyed and drying at the dyer's. A light dress of persimmon color

I'll dye the summer dress for you a light brown

At the dyer's in Kyoto I learned about dyeing—patterns and how to attach the collar

It's easy to dye and match the stripes

Fitted perfectly together, three stripes along the lapel

This morning on the way to the hunt he dropped his extra coil of bow-string 21

Bow-string meanders. Absent-minded he dropped his extra coil of string and paper

Confess. Whose string has fallen here

I've scraped up pepper, lots to fill the pepper bag

Take a pinch and try. Pepper's a bitter thing

22 We did our planting again this morning. Seeds for flax

Whose shuttle, I wonder, shall I borrow

The one I came off with was the eight-shuttle from my sister's place

When the leaf of the flax is triple, pound and peel the bark

Heating and peeling, pounding and peeling, my busy hands

23 A pheasant cried this morning, Love, at the first sign of light

A pheasant or a falcon cried passing on the wing

Fool of a pheasant to cry fluttering his wings

Fool to cry, so the falcon eats him up

Fluttering wings and then a long fall deep into the mountains

The sound of the bird's cry this morning was a good sign 24

'Forty-five bushels of rice to a half acre.'
That was the fair sound of the bird's call

A good bird sang: 'Forty bushels of rice'

A cock sang deep in the night so my lover went home

Listen. The fine song of a bird, a lucky sound

The morning's work moves smartly, girls, planting seedlings 25

The barns will brim at autumn harvest

Fast ripening rice. Plant it. Before your eyes it twists and twines

Waving rice smoothly ripples towards the storehouse

The barn is piled high. Make this year's harvest into the seed of next year's crop

26 A fine sight. The grand hunt at Fuji

All in order—bows, falcons—the grand hunt at Fuji

Beaters at the grand hunt at Fuji—horsemen, fifty thousand strong

I myself took part. . . . . . straw apron for Daisan Fuji

Won't you lend it to me. I'm on my way to the shogun's hunting ground

Shaggy apron. What they wear at the hunt is splendid

27 The middle of a green moor is where deer crouch low

Set bows in order. Get arrows ready. Deer crouching low on the green moor

If you want, shoot, then. Deer crouching low on the green moor

I was happy this morning going out the gate. First call of the deer

Have they brought lunch yet, girls? What are they carrying on their shoulders, I wonder 28

Seaweed from the beach at Wakasa is what they are bringing out

They cut and piled up seaweed. The priest had it with his meal

Let's cut it for our lunch. Seaweed hidden in rock crevices

They are carrying food on their shoulders. Men are on their way now

Eight tubs of rice, girls. Nine rice scoops 29

. . . . . Chiyo, ...

. . . . . piled up ninefold

Ask Chiyo for what you want. Chiyo is in charge

Over the rice tub set a gold scoop upright

For the workers in the fields today they prepared a thousand bushels of rice

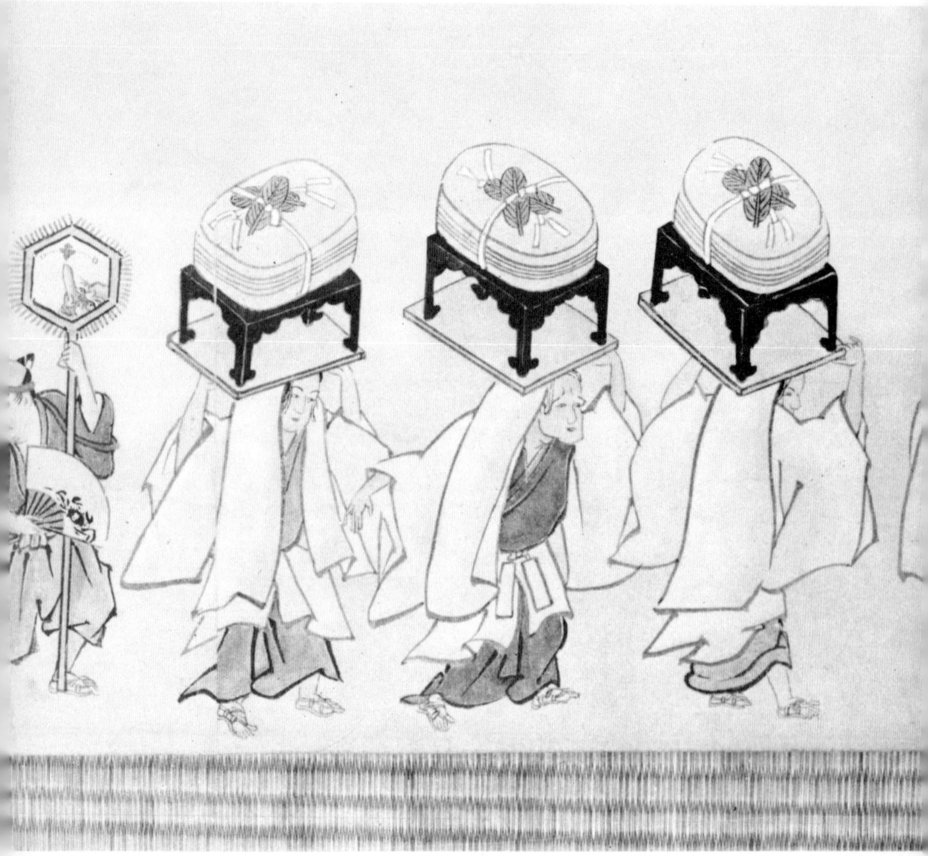

First Set of Noon Songs

He whittled the dragon's white mustache into a toothpick and put it between his teeth 30

He finished lunch then put aside the fitted box in the middle

The owner of the field we plant today put a gold toothpick between his teeth

For making toothpicks, the loquat tree of southern India

I'll give you a toothpick dyed with the May flower

Women, if you dye a toothpick with flowers it's low class

Sad, I dropped my favorite toothpick

How far should we escort the women who brought the food before bidding them farewell? To the mountain barrier 31

Mountain barrier. Barrier at the temple.
Port called Tsumi. Last port on the way

We have all but seen them off. Look at that. Man-made island at Hyōgo

Hurry, slow one. The tea-house at the pass is near

32 When lunch is over there is plenty of clear wine

It's fine, clear wine that makes a voice sound good

The voice sounds good with fine, clear wine to drink

When you drink wine, lees go down and you find yourself singing a song

The man I like is good at making songs and poems

33 What flower blooms in the field behind the landowner's house

Flower of the mulberry, flower for wine, flower of wealth

In the field behind the landowner's house blooms the flower 'money'

What sounds in the storehouse is money

So much money in his storehouse, that they had to heap what couldn't get in into wagons

Make money. Money is what gets things done

If you want coins, the guild of horse traders has them

Flowers on the chestnut tree bloomed thick 34

Will there be fruit or not? Ask the flowers

Flowers on the chestnut tree in bloom made bright a whole hillside

Flowers bloom and on a branch bent heavy fruit appears

It must be Kyoto. It must be Nara. It must be the temple Kōshōji 35

Where the lotus bloomed. It must be the temple Kōshōji

If it rains, I'll ask to use you as an umbrella. I'm speaking to you, lotus leaf on the little pond

Terrifying. Lotus leaves coiling round the lake

36 What blooms in this field is the flower of the creeper

Take plenty in your hands and we'll set off for the grand estate

When you go to the grand estate have hollyhock in your hands

Be sure to gather flowers before we set off for the guest room at the grand estate

Second Set of Noon Songs

Hurry, weaver, weave summer things 37

Cut, tailored and woven things . . . . .
coarse-grained cloth

Let me dye the summer dress with its fine close
weave

Let me dye with flowers the summer dress for you

Ribs for a fan, girls. Cutting out seven 38

Cutting out eight. With a tap-tap let's
hammer the rivet home

Ask for it and I'll fold up the fan for you

Maybe a little senile so I miscounted, calling seven
eight

I'm going on a pilgrimage to Utsukushima. On
the way back I'll give you a fan

39 When you go to the Capital, in the forest near Muronotsu

Crying bulbul, for whose love do you cry

The bulbul is the bird that cries: *hiyo-hiyo*. The mandarin duck is the bird living by the pond

Mandarin ducks on the pond never taking flight with the one they love

Tail feathers. I have one from the mandarin duck—something to rely on

40 Deeper than the shadow of a fan is the shadow cast by bamboo

What really feels cool is the shadow cast by bamboo

The men didn't shade themselves. But the young man covered his head with a fan

It's better to put up the umbrella. What's this about shade from branches

If it blows it will cool off, won't it? Blow wind

When you go to the Capital, first of all, in the lodging house at the port 41

Shooting at the target, aiming at the stuffed bull's-eye, he hit the mark

The arrow he used for target practice had feathers of a white heron from the Capital

In the feathered quiver thrust thirty-five arrows for target practice

Buy. Try your luck. One try—thirty-five *mon*

A single fan. Two women 42

Tear it. Tear it apart. Tear it in the middle

Try tearing. The fan can't be torn. The girl I love is faithful

I'll give it to the girl. The fan painted with Kyoto pictures

43

When I got up this morning and looked out at the temple across the way

Beautiful boys had broken off flowers to put in their hair

Let us go with flowers to the temple at the grand estate

Let us go with flowers. Music is starting now

I can't break the spell of the music

Songs When Wine Arrives

Wine is on its way. What shall we eat with it? Endive 44

Soak endive in vinegar, a marinade to
have with wine

With wine endive growing in the flax field

When you go to Gojō they serve endive with your
wine

Let us pour out wine for the field workers today 45

Come, let us pour it out in happy cups

Have some to drink. It will make you young
again, best wine from the bottom of the jug

Let them serve you wine, sweet chrysanthemum
wine from Kaga

And with it salted beans, first of the season from
the Capital

46 Let us offer wine to the field god, fine wine

Into a wine ladle with a long handle, into happy cups

An offering of wine to the field god is made with three dips from a dipper with a long handle

An offering of wine to the field god is leading water into a sandy field

47 'A speck of dust. A speck of dust.' A speck of dust is in the wine

If there is dust don't drink. We are going to dance

You drink but never get drunk. Isn't that the way with what you buy

Cheap wine. For five *mon* eight containers full

It is the usual thing for wine made at harvest to be watery

They say: 'Drink, drink,' and make you. Isn't that the way with drinking in the fields

After Drinking

They drink wine. Then the field workers really set to work 48

They drink wine. Then they can say whatever unpleasant
things they want to

Sing a song when the field god's wine has made you drunk

Kite, kite, dance high in the sky. We'll thrust a roasted rat
up for you to reach 49

The women are fine dancers

Let's learn a step, a dance from the repertory of
the Kongō Noh school

I danced a dance but I forgot the rest of the song

50 'Ready your bow. Ready your arrows.' We did get them ready, still . . .

Tarō-Tarō wanted to see the Capital

Let's take him with us and worship at the Kiyomizu temple in the Capital

I'll sit down to rest beside the pine called Sango and compose a poem

Larch at the Capital were very much like cedar in the country

Third Set of Noon Songs

The cuckoo. Tipping her field hat to one side 51

You hear it but never meet it. The cuckoo

She tips the field hat she wears in the fifth month towards the one she loves

The rush for her hat is ten-strand. The band for her hat is a cord made of eight-strand Mishima cotton

For the hat made at Daisan, a cord of twill

The umbrella that passed by yesterday is on its way again today 52

If it's going by let it go by. I wonder if anyone will have him as a son-in-law

It's settled, though no one knows yet: the Gion priest will be my son-in-law

Let's make him the groom. The priest who is down from the east

Let people say what they will, I won't take the priest as my son-in-law

That umbrella over there, what umbrella is it 53

An umbrella for May rains? An umbrella
for love?

To hide himself he put up an umbrella for May
rains

Are the decorations on the umbrella set or ones
the owner liked

Decorations on the umbrella. By all means let us
ornament it

Boys and girls, look. Decorations on the umbrella

Umbrella in a narrow lane 54

Unfurled, opened out, put up it can go by

Fold it up and pass by with it over your shoulder,
priest

Go by with it up. How fine a thing to look at that
way

55 Yesterday it came down from the Capital, the white field hat

Your're going to lend it to us, aren't you, the white field hat

Give in and you can wear the peaked hat from Kyoto

Give in and you can put on this pretty hat woven with fine rush

Don't let young people punt over the flooded fields without hats

56 Reeds cut deep in the mountains. Were they cut yesterday? Were they cut today

What was cut yesterday withers. But what's cut today, what's cut today

Withers, but look at the leaves of what's been cut today

Cut what has been set upright. Thatch for the eaves

57

They say there is a dance for the god at the Kasuga Shrine

Let's be on our way. They say that there
is a dance for the god

He wears a fleecy head-piece dyed red, the director
of music at the Kasuga Shrine

He wears a fleecy head-piece dyed red and I love
the one who directs the music

Put it on. The fleecy head-piece makes a mighty
noise rustling

Fourth Set of Noon Songs

Rice in the field before the rich lord's gate 58

However much you harvest, it never diminishes. Rice in the field before his gate

However much you harvest, it never diminishes. Early ripening rice in the field beyond us

If you don't press the rice stalk down you won't be able to thresh it

What has the cuckoo brought 59

A dry measure, the strickle and bags of rice

It brought bags of rice to store in the north-west corner of the barn

Take what you want, they never diminish. Bags of rice stored in the north-west corner

There is lots of rice so weave plenty of straw bags

▲小春めでたき。百姓ぐるみ内に花さく▲金乃蚕の。ちからねは俵よあまる嵐
はてへをかけてまふうたふあり翁とんをこがねのまかちらしねはとうまふ俵よ。つみるこからしともゑくしくいくなびもう幸よ

▲ほとゝぎとゝり
そへ子夜泣をとい
これよ

▲郭公乃ちらく
まくにあ紙も
ゆすられ

▲時鳥の声より
あをたりしろき
早乙女

▲後のろふかせとも
らぬるよふなく
通ふ稲妻

▲彼ぶらしやうふは
きやげふうくし
く光る君

60 When they say *uguisu* they mean an elegant mountain bird

An elegant bird, concealing itself in the mountains to compose poems

When your voice becomes hoarse, borrow the *uguisu's*

Sing out loud and clear, *uguisu* of the fields

One never tires of hearing a fine, clear voice

61 Look at the streets in the Capital. Broad avenues

Seven lined up, broad avenues

Marketing has begun. Let us sell the set of seven fitted bowls

Taking the stranger along, setting out for the Gion quarter

Taking the stranger along let's be on our way . . . . .

Dressed in our best, at the town Saijo let us sell . . . . .

Going up stream, going down stream, sweetfish travel in threes 62

Into the rapids with you. They live in the rapids and never go into the shallows

Don't muddy it. Up stream where they catch sweetfish

Set up fish weir. Down stream where they catch sweetfish

Unsalted dried fish. She's so lovely I don't know where I'm at

The youngest daughter spilled wheat in the nursery-field, so it seems 63

She really did spill it, so it seems. Shook and dumped it out, it seems

Now is the time for fondling Satsuma wheat

Go ahead, fondle the wheat. Women at the shrine

64 Yesterday from the Capital came rice called *mekuro*

Three handfulls yield forty bushels of rice

Seeds of prosperity. Sow three *go*, harvest fifteen bushels

Double-headed stalks spring up. They do on this thousand-stalk plant

Scatter seeds of good fortune

65 When the swallow has all its feathers it sets off for a far place

It's sad for the parents who raised it

With all their feathers they set off for Paradise

Don't let the parents of the swallow feel bad

They had their feathers so all set off together

Beach over there, small beach and tiny white waves 66

Kicked. The ball rolled over and over

Fine pebbled beach. I thought, 'Small rocks. I'm in love'

Hold your skirts high. There are big waves along the strand

It was love and seaweed I hurried after along the beach. I got a clam

If you look at the picture on the fan, look at the one on the other side 67

Look how the six year-old writes his childish letters

Letter writing. Copy-book specimens like grown-ups write

With fine small letters you will get better and better

Traveler, look at the back-side of the fan

68 Come on, come on, time to go back. Look at today's sun

The sun is on its way down. Look at today's sun

I forgot and left my plaited hat at the teahouse. I dropped my fan at the market

I'll buy one to take with me. I'll buy it at the next town I reach, Miyoshino

There was none at the market so he won't come back with a fan

Summer is passing. Let's return the fan

69 The visitors are here, girls, wearing white summer dresses

Fine summer dresses. The skirts are of patterned cloth

They get dirty easily. The skirts of their dresses

It should be fine. I had to pay five *mon* for the dress

Even though it is dirty, praise it. The dress has a white ground

First Set of Evening Songs

Let us make wine, deep clear wine, from early-ripening rice 70

Let us make wine with clear water from the spring at the foot of the willow

How far should we escort the women who brought the food before bidding them farewell? To the man-made island 71

To the man-made island. Because of my love . . . . .

We have seen them off. Look at that. Man-made island at Hyōgo

The interesting places are Hyōgo and Nishinomiya

What I like to lie beside. The body of the young girl. It is soft firm flesh

Say what you will, I never get tired of that young girl's body

Sleep with her one night and you're mixed up seven days after. The body of the young girl is beautiful

72 The officer on his way from the Capital. How has he covered the pack on his back

Cover of brocade tarpaulin. Skin of a fawn. That is how he covered it

The pack on the man's back. He covered it with white skin from Shirakawa in the Capital

A light garment with a pattern of young deer, not offering her body as a usual thing

73 Down from the Capital. A carpenter from the Capital with tie-dye patterns next to his flesh

On the shoulder a printed design of bows and arrows. At the waist a printed design of quivers

A thin summer dress with tie-dye patterns. When will we dye it, fall in love and have each other

My lover left the Capital yesterday 74

I expect him to arrive today by the road along the river bank through Harima

The highway through Harima. There's a short cut

But to go the long way round by the Harima road. He can't be thinking much of me

Women from the Capital. What is on their skirt 75

Ink, inkstone, brush and a deep crimson *obi*

Ink and inkstone. Ask and see. She doesn't have anything at all

She had an inkstone, an inkstone to use for writing

76 My lover serves in the Capital cutting grass, they say

Sweet rush, spur grass, reeds. They say he is cutting grass

My lover was commandeered for the labor force to cut grass

Without a scythe to cut with he'll probably wander lost in the mountains

77 I put a piece down on the chess board but know nothing about the game itself

I shake dice for a twelve-spot but know nothing about the game itself

My lover is expert at backgammon

As good as that. Then, bring the board over. I'll roll the dice

The interesting game is backgammon

What a sight! Merchants from the Capital 78

Carrying chests of merchandise on their back, traveling three together

In their chests are many, many treasures

With treasures on his back he came from the Capital today

Down from the Capital with more presents than I expected

Do I love the merchant? Do I love his chests 79

It is the stuffs dyed deep-purple in his chests that I love

In his chests light dresses covered with designs of the plantain tree

I've been bewitched by the deep-purple

Let's try it on. The light dress made of threads finely twined together

80 Warriors from the east are expert bowmen

He shot down a bird on the wing as it darted into the sky

Yes, expert. He brought down a crane dancing high into the air

Bow hand of the young man who struck a bird just darting off

He shot it. He shot it. Wonderful how he handles the bow

81 Oh that I had a bow and arrows for the east

I am on my lord's service setting out east

'You like a warrior's duties,' is what people will think, won't they

Service in a samurai family is quite a spectacle

I'll give it to my lover. The fine lacquered bow wrapped round with wisteria vine

Bow that I love. Girls, if you pull . . . . . come, come 82

If it's the bow-string I made myself and you pull . . . . . come, come

My lover's wrapped bow. Want to know how strong it is? Pull and see

Let's give it to the bridegroom. The bow wrapped round with wisteria vine

When you're in love, even without pulling. Come. Bow-string of Seki

Planting women in the fifth month and the spring warbler 83

No place it hasn't been to already. Warbler in spring

Let's hear your piping voice, warbler in the basket cage

A bird now, no longer in the cage in the guest room

An empty cage hangs down in the guest room

84 The lord's horse, the lord's horse, where is it tied

Beyond the ridge, beyond the valley, tied to a bending pine

Beyond the ridge tied to a pine at the edge of the field

Tied on the moor. Let us get it going

85 A priest came from the east. He came down alone and one side of his face was sunburnt

Let me have the priest. I want to make him my own priest

Stay, weary priest of the road

A strange kind of priest who doesn't read the Lotus Sutra

I snatched and knocked it off. The round rim of his tall plaited hat

The priest on the road took his lodging for the night while the sun was high overhead

That rise in the field beyond, what hillock is it 86

It is the blessed . . . . . hillock of the owner of today's fields

Let us set off, then. If we go he will give us a blessing

Ah, how lucky. I went and brought fortune back with me

She rolled into the newly ploughed field. She dropped her hair band 87

Tear a piece of paper for me. I want to use it to bind up my hair

My *obi* will keep me true. The *gochō* paper will be the band for my hair

No, I won't untie my *obi*. Taking it off means you're very serious

On his way back from the Capital he brought me an *obi* I hadn't expected

88 'Long sword for sale. Long sword for sale.' Long sword for sale—but no sheath

Here is a sheath for you, one made of bear skin

Look the sword over carefully before buying. The fine, thin-bladed Moriyama sword

Someone carries the sword on ahead for him

When the young man draws his sword and stirkes his pose, tip your field hat

89 Radiant bird. Bird of fine color

Mouth—the gold and silver of brocade . . . . . A white bird with wings the color of the lotus

Wings are white. Isn't that the way with birds in the Captial

A bird from the Capital? Sing. I want to hear your voice

The color of the bird's wings. The first time I've ever seen wings that color

Atop the mast of the Tsukushi boat a stately bird 90

Beak—silver. Wings—white. Legs . . . . .

White, black, the bird's five-tinted wings

The shipmaster raises a marvellous bird

A bird on board. I suppose that is the custom in the land of Tsukushi

Lord Hatakeyama Rokurō carries a short sword at his side 91

It cost about a thousand *ryō* of gold

The maker of the short sword, Niō Saburō. The maker of the long sword, Kanemitsu of Bizen

The short sword Rokurō carries is the famous one that pierced the Ainu

92 He wears a long sword. The blade of silver and gold

It hangs low, slung at his side

The sword Minamoto wear is made of gold

I would like to carry a gold sword but I don't have the rank

If you are looking for a sword to buy, buy one made of gold

93 Plant your favorite willow in the field before the gate

Its branches will flourish in the field before the gate

I won't plant willow. Willow brings bad luck because its branches hang down

The bending of the grove groaned toward Yoshino

In the darkness of the grove let us meet

Seeking the odor of the plum the warbler passes from twig to twig 94

Well anyway, gently, warbler. Don't spill the flowers, warbler

Pass along the branch, then, but don't spill the flowers, warbler

Let's hear your piping voice, warbler in the cage in my hand

The warbler chose the plum on which to rest his wings

Rice-planting girls in the fifth month eat plums, don't they 95

Plums are sour. They set the teeth on edge. Peaches are bitter

I ate one of the crimson plums which Sugawara Michizane liked

Have a plum. With a plum your voice improves

Have one of the crimson plums which Michizane missed

I'm planting it for a present, the Chinese plum on the bank beyond

96

The bamboo sticks from Kyoto, girls. Without breeding you can't play them

The two prostitutes? The young man, then, I suppose?

No one in the world can play two sets of bamboo sticks at one time

I'm telling the truth: no one can play two sets of bamboo sticks, can they

Someone who's good at playing two sets of bamboo sticks

Second Set of Evening Songs

Along the banks in Harima they strung up bird-clappers 97

Let the man from the Capital pull as he passes

For her daughter the clapper in the rice field. For her daughter-in-law the clapper in the millet field

Clapper in the millet field. Without even pulling it sounds: *danko*

Clapper in the millet field. Even when you pull, it doesn't sound

Look at the footpath. See her fluttering her sleeves 98

Don't flirt there along the footpath

Come out and see what it's like to be a coquette

Blow something, a tune for planting, the one called '*aigyō* seedlings'

They may be girls, they may be young but don't let them flutter their sleeves

99 In the rapids where rivers meet the boy dropped his flute

Set up the weir. Set up the weir. The flute will come to rest in its net

Don't blow the bamboo flute of love. The sound terrifies me

When the boy stole in to love he blew his flute

100 Do a good job planting the nursery field. My daughter-in-law is a robber

The person who said that has a young daughter who makes off with rice laid out beside the pillow of the dead

At any rate, that's too bad. Your daughter-in-law is getting a name for something she never did

She may try but can she deny she's slandering her daughter-in-law

I wonder what's sprouting in the mother-in-law's mouth

Preparing a field on the hillside is quite a job 101

The monkey plays the bamboo sticks.
The badger strikes the drum, they say

When he strikes, it sounds pretty good. I like the badger's drumming

From ages back the bamboo sticks are what the monkey plays best

How long will scarecrows stand in the field on the hillside

Fast pony there. Look at the frantic baton 102

It champs the bit. Look at the frantic baton

Shirō is the man for mastering a horse

It's the spirited horse that Shirō leaped best upon to still

I think Master Shirō could mount even the dragon of King Tosotsuten

103 Don't make love with the man who's beating the drum,
patch the place on your shoulder and wear it

If you want to tear, go ahead tear, though
I love him very much

Not the patch on the skirt. The main thing is the
patch on the shoulder

The colored patch on the shoulder. It's not as big
as I thought

It's too bad when the one you love is out of your
class

104 Better than an acre out there is a half acre by the gate

. . . . . a half acre field by the gate

If you can't give that, then give me a portion of
field out there

. . . . .

Who built the temple in the mountains over there **105**

Emperor Kammu. Dengyō-daishi

Better to see than just to hear of. The temple over there

Get dressed up. We'll set off for the temple in the distance

Who built it? The Yakushi hall in the mountains over there

Third Set of Evening Songs

106 Young men three together. I wonder which is our son-in-law

Wearing vermillion dress, carrying a decorated pole in his hand

Carrying a fan in his hand, sitting formally in the guest's place—the bridegroom

Play one match, Sir. The game with decorated hairpins set upright

They are on end. Play, Love

107 Lady, when you want to know his rank, look where his hair is bound

Inserting decorated pins, tying back the hair with a . . . . .

Where the hair is bound back, certainly looks jet black as if ink had been poured over it

Let me bind it back. Red plum colored . . . . . to bind the hair

108 Young men contesting and dark blue when wet

How beautiful to look at dark blue when wet

Red plum wet is a fair sight. Girls wet is not so good

I can't make her mine—red plum. Probably because I wasn't meant for her in the first place

Have the girls wear a dress dyed red

109 I suppose I should dry it by night. Fabric dyed at Harima

A design of all kinds of chrysanthemums. It will turn a fine color

I couldn't dye it after all. So at dawn I scattered real chrysanthemums

I learned the trick of dyeing by patterns at a dyer's in Kyoto

A plain dark colored apron for the dyer's daughter-in-law or for his daughter

Pretend you are a bow. Black lacquered and wound tight with wisteria vine 110

I wish I could lie close across the shoulder of the warrior from the east

Pretend you are a bow. I'm in love with the soldier from the east

Try as I may I can't get close to my love. Poor me

Beautiful. The way the soldier from the east looks

I passed through a grove of chestnut trees. Chestnuts were falling: *tei* 111

In a torn sleeve there is no staying

One fell. An early chestnut on the magic mount Sayoutsu

Falling fast now. That's the way with early-ripening chestnuts

Shall we be on our way. Shall we go. To the distant grove of chestnuts in the mountains over there

112 Below the plum tree he kicked the ball: *to*

Plum blossoms spilled. Ball stayed in the sky

He kicked it up. He's good at kicking the ball

In his pale blue dress he kicked the ball. I love him

My lover is good at ball

113 Willow at the gate of the lord's estate. I'd like to see the ball get caught there

When you pass by kick and be gone. I'd like to see the ball caught there

Have some tea. You kicked the ball so your throat is dry

He kicked the ball. He doesn't take the tea cup. He takes her hand

Kick the ball before you pass, Sir

They dusted the guest room. They spread mats 114

Come this way, please. Won't you take your seat over here

This way, this way. Toward mats edged in a pattern of cloud and chrysanthemum

Can anyone match that? The way the one sitting squarely there is dressed

Riding a horse, three men. I wonder which is the bridegroom 115

Next to his flesh white weave, purple weave. A short-sword hanging at his waist. The bridegroom

Wearing a short-sword, riding in the middle. That's the bridegroom

Bright. Bright. The costume the bridegroom wears

116 I'll put Master Tachikuse into a bow wisteria bound

I want to enclose and wrap him in. I'll put him into a bow wisteria bound

I'm beside myself because of Tachikuse, in the beating of snipe's wings

Whatever happens, I can't forget Master Tachikuse

I'm confused. I thought it was Master Tachikuse

117 You can't tell a real rice-planting girl until the middle of the night

It's only in the middle of the night that her voice becomes really good

When the sun goes down the lovely voice of a planting girl

Planting girl of the fifth month, voice at sundown is fine

Her voice is lovely. I'd exchange it for the short sword hanging at my belt

Lovely. Sing out, planting girls in all the fields

Three riding a horse. I wonder which is Manjurō **118**

Dark blue reins. Whip. On the horse the one in the middle is Manjurō

Boys getting together. Dappled horse with pink reins

Can anyone match it: how he looks mounted on the horse

Boys together.

Fourth Set of Evening Songs

Lady Tora of Oiso is drowning in love 119

Lord Soga Jūrō comes along to ease his passion

Lady Tora, enough love making for now

In love you make up songs

Come tell me, Lady Tora, when did you see me and fall in love

The white sleeve has picked up a scent of musk 120

I slept with Lady Kuruhara. The scent came off

Musk, perfumes adhere. The whole sleeve smells

Go ahead say what you will. I'm not likely to forget Lady Kuruhara

Never ever to forget that look, her hair still matted from our sleep together

121 Shells stuck fast to white rocks along a distant shore

Pebbles. Because I was in a hurry? They clung tight

Hey, over there. Adorable dimpled face

Little dimpled face sucking fast its whistle

Nothing wrong with being bald. You don't need tweasers

122 Could it be my lover coming? The rear door's creaking

Kiriri kitto, it goes. The rear door

Maybe my beloved. Maybe a puppy howling, can't get the door ajar

The short sword put by the pillow. The long sword against the screen where it folds

I want to find you. Which room are you sleeping in

Fine Chinese thread is what they're rolling up 123

If Tarō comes along, quick reel him in too. Tight

Chinese thread going round and round, *kara-kara*. Wrap it tight

Meeting you . . . a tugging at my sleeve

There's nothing better than Tarō's just being here beside me

When thread gets tangled, you can set it free again, yet . . . 124

People tangled together. It's not so easy to untangle that

Not so easy to untangle. Lovers enmeshed

A pair of wine bottles set side by side. Intimacy

Too much intimacy and a third party's interest cools off rapidly

にもうさと称ふ
▲家に住つ。冬雄は。
早苗とられていゝ
かゐせん
▲さなへにすかる。
いなごも。とりれ
植田に紛よ
田鳴は悲くぞ
かゝりふしまする
里くゆかる声
し大きさはこの
嬉うなり
▲年ぬちし光の。
わづらひも。田
作どまのき

田植

妻子のとも下女
けしも田うへをする
拍子そろへれく
みうたうたへあこ
ゐりて今ゆへ
風流にてうたふ
植るなりけり今乃
嫁むすめをもと
ひて催ける婦
よめをやとひの
たくひはなきさ
しくすなわち
まいその日う
ゆう子苗をとく

125 Break off the lovely branch of the cherry and come

To decorate the room we'll sleep in.
Break it off and come

Flowers. Why? A long sword's the thing to decorate our room

To adorn the room where we sleep, bring flowers from the distant hills

126 After sundown wear a thin white summer dress

Women of pleasure, rice-planting
women, wear your best this evening

Thin white summer dress. I'm beside myself with these girls

I like it. Talking to girls.

That thicket in the distance is a tower perhaps? Or a pleasure chamber 127

Neither tower nor pleasure chamber. Still, a thicket is a fine place to lie in

Better than mats. A thicket like that is the place for lying

The best part of sleep on the grass is doing it two by two

I brought down the one I loved in that bamboo grove once

Sun goes down. Against it two snipe fly westward 128

There's a lake in the west too, they say

Along the pond deep in the hills into which the snipe plummeted down

Are birds making love? Their cries are pitched so high

Two, paired. Close-set. That's what it is to be together

129 The sun's going down, girls, upon the edge of the western hills

A lotus flower bloomed, bent over and hung heavy down

Bloomed. So it will drop, I guess. The lotus in a temple among western hills

Today's Lord Sun soon to touch the edge of the hill

130 Paradise in twelve directions, girls. We planted a tree

Was it pine? or willow? The tree we planted

Within thick pine, upon the dark plain of Yawata

A cuckoo rests upon the pine. A warbler upon the dark bamboo

Dark. So a carpenter lights his shavings

Someone's painted a round moon, they say, upon a sky-dark cypress fan 131

If my lord really painted it, he's a man of feeling

He painted it, for sure. A cypress fan upon a white-ground screen

Beautiful she may be, but without touching her nakedness she remains an elegant lady on a painted screen

The passionate lover slipped a fan he liked into his belt

The master pilot of the boat for Tsukuji mumbles in his sleep 132

Trays: Seven. Tea cups from Tenmoku: Eight. And coins: Sixty *chō*

The bridegroom. The bridegroom. On his way now. He opens out an umbrella. He works his oar

Come out, look. The bridegroom. There's no mistake

You'd know him by the heaps of treasure he's piled up.

Song to Mark the Close

The lord of today's field has planted them in full

He's put up eight storehouses in a row. He's opened the doors wide to fortune

We've made fat ears of rice nod. Let's put up storehouses. North, South, East and West

A real Kyoto locksmith is making the lock for the storehouse

The lord of today's field has earned his name: 'Doubly, doubly rich'

Let water run full in the river for our washing

Notes

Stanza

5 '*Ura-ura* he cries on the wing.' This and other onomatopoetic expressions are retained in translation where the context itself explains their meaning.

6 'He took a tuft from the hairy belly of a deer . . . then used it up to write out the Lotus Sutra.' Humor and abrupt transitions are elements in the original.

8 'See how the officer of the fifth month rowed deep into night.' The sound of the words is more important here, as it is in some other places, than their meaning. In the original an s-sound is repeated. This use of the s-sound may have had a special importance in the ritual since many of the words related to rice-culture contain it: for example, *saotome, sanbaisan.*

9 'The little one I've brought along with me hasn't yet got used to people.' Probably meant to be taken with two meanings: the 'little one' is both the young of the warbler and a young planting girl coming into the field for the first time. Such doubling of meaning is an important element of the poetics

of this sequence. For example, in #11 the newly ploughed field being raked by workers is spoken of as a salt flat by the sea.

14 Talk about dyeing cloth seems to be beside the point in a rice-planting ritual. Actually it is appropriate for several reasons. As explained in the Preface, clothes which the girls and boys wear in the field constitute an element of interest among songs for this time of day. In fact, a part of the ritual is for young girls to wear their best cloths even though they become muddied by the end of the day.

17 'Serving at court keeps you busy. Next to the paper partition.' The traditional way of separating area from area in Japanese homes is by the use of paper partitions.

18 'The one beating the drum is Kudotarō . . . Our own local Tarōs and Jirōs will have to fix themselves up too.' Kudotarō and Kajiwara are names given by local tradition to the musicians who played for Shizuka Gozen's dance as recorded in the warrior tale called the *Gikeiki*. The tradition in the accepted text gives their names as Kajiwara Kagetoki and Kudō Suketsune. See Helen Craig McCullough, *Yoshitsune: a 15th Century Japanese Chronicle*, Tokyo, 1966, p. 232 f. Tarō and Jirō are simply common names for local people.

21 'This morning on the way to the hunt he dropped his extra coil of bow-string.' An extra coil of string was customarily carried on a spool. We may imagine some sort of dramatic encounter implied in the song. Early in the morning the man goes off in a daze after a night of sex. Her girl friends scold the girl afterwards by asking about the momento he left behind. Finally, it all turns to bitterness for her as she tries to conceal her emotion through a word association: paper and pepper.

22 Here the doubling up of meaning is not simply to bring about an association with the clothes the girls and boys are wearing, as explained in the note to stanza 14. A completed natural process in one area—flax into cloth—is a good omen or presages the same for the rice-seedlings, i.e., a large harvest, eventual completeness.

24 " 'Forty-five bushels of rice to a half acre.' " Far beyond the normal yield. This stanza seems to enact an opposition between the traditional aristocratic culture, within which a bird call early in the morning dispatched the man home from sex and so was an unwelcome sound, and an agricultural tradition

in which a bird call was a propitious sound or omen.

26 'A fine sight. The grand hunt at Fuji.' Refers to the famous hunt around Mount Fuji which is the climax of an oral tale called the *Soga Monogatari*. Daisan Fuji may perhaps be the mountain of that name in Tottori Prefecture or a case where the name of the majestic Mount Fuji near contemporary Tokyo has been applied by analogy to the largest single mountain in any other area, in other words, any prominent local hill. Implied in this stanza is the gala spectacle of the transplanting day itself.

29 '. . . . . Chiyo . . . . .' One of the most puzzling verses. The context seems to require that Chiyo be interpreted as the name either of a particular woman or a role whose function is to manage the ritual meal at noon. Perhaps it applies to one of the *unari* in stanza 31 and explained in the preface.

31 'How far should we escort the women who brought the food.' This refers to the *unari*, and the phrase might equally well be translated as 'visitors'. One role these women filled was to bring food into the fields for the workers. There is a doubling up of meaning in this stanza: images of traveling are applied to the rate of work in the field. Inexperienced girls are slow.

'Look at that. Man-made island at Hyōgo.' Another case of using an allusion to a recited tale, here the *Heike Monogatari*.

34 One of the principles which make of these songs a single long poem can be seen if this stanza is compared to a song from the late afternoon, stanza 111, where the chestnut trees which are only coming into flower here finally bear fruit.

35 The use of Buddhist and non-Buddhist imagery side by side in stanzas 129 and 130 seems to come from much the same spirit which allows the lotus to bloom here in company with natural field flowers, all as an image, and prayer, for the rice crop.

40 The heat of the day is being felt in the fields by the time this song is sung.

44 'When you go to Gojō they serve endive with your wine.' Gojō is a place in Kyoto. Again the culture of the countryside merges with that of the Capital.

59 'It brought bags of rice to store in the north-west corner of the barn.' A magical place.

62 'Unsalted dried fish. She's so lovely I don't know where I'm at.' Some of the subtlety of association in these poems can be seen here where a word pun allows a transition from fish to the attractiveness of a young girl. The next stanza develops the idea into a sexual meaning. Fondling wheat seems to have that sort of connotation, as does 'clam' in stanza 66.

64 'Yesterday from the Capital came rice called *mekuro*.' An especially fine type of rice seed.

65 This stanza refers also to the planting girls and the young men in the fields.

75 'Deep crimson *obi*.' *Obi* is the name given to the broad waist band used to bind traditional female dress. Here the *obi* is evidently only one item of a pattern printed on material.

82 'Bow-string of Seki.' A product of a place called Seki in the district of Ise.

87 'The *gochō* paper will be the band for my hair.' One explanation is that the word *gochō* developed from the expression 'small square paper.' In stanza 21 the word was used because of its similarity in sound to the word for 'pepper.'

90 'Atop the mast of the Tsukushi boat.' This refers to boats based around Tsukushi, an older name for Kyushu or a district of it, and operating in the Inland Sea.

91 'Lord Hatakeyama Rokurō carries a short sword at his side.' A samurai customarily carried two swords. Hatakeyama is the name of a famous family in the Japanese middle ages.

'The maker of the short sword, Niō Saburō . . . Kanemitsu of Bizen.' Names of actual swordsmiths.

92 'The sword Minamoto wear is made of gold.' The famous Minamoto, or Genji, family controlled Japan for a long period.

95 'I ate one of the crimson plums which Sugawara Michizane liked.' Sugawara Michizane was an historical person of the Heian period who became a popular legendary figure.

96 'The bamboo sticks from Kyoto, girls.' One type of these sticks, called *sasara*, are flared bamboo sticks struck together by the leader of the transplanting ritual to accompany his singing.

97 'For her daughter the clapper in the rice field.' A traditional theme in transplanting songs is the mother-in-law who dis-

likes her daughter-in-law. Here the older woman makes the younger chase birds out of a field where the less valuable grain, millet, grows while her own daughter is given the job in the better field.

98 "Blow something, a tune for planting, the one called '*aigyō* seedlings.' " As well as the name of an actual melody performed by the musicians, '*aigyō* seedlings' seems to be used in part as a pun on 'coquette.' The fluttering of sleeves seems to refer indirectly to the movements of the planting girls in the fields as they work. It is bewitching.

100 'My daughter-in-law is a robber.' See note on stanza 97. This stanza seems to suggest, as do others, a sort of dramatic dialogue involving a critic of the mother-in-law's severity in her treatment of the younger girl.

102 'Shirō is the man for mastering a horse . . . the dragon of King Tosotsuten.' Nitanno*shirō* is the popular name by which a warrior in the service of Minamoto Yoritomo became known in legend.

King Tosotsuten is a figure in Buddhist mythology regarded as a defender of the faith.

111 'Chestnuts were falling: *tei*.' *Tei* is an onomatopoetic sound for the falling chestnut no longer in current Japanese.

119 'Lady Tora of Oiso is drowning in love. Lord Soga Jūrō comes along to ease his passion.' Characters famous in the *Soga Monogatari*.

132 'Tea cups from Tenmoku.' A place in China which gave its name to a special kind of ware.

A Note Concerning the Text

by Manabe Masahiro

THE EXPRESSION *tauezōshi*, literally 'bound book for transplanting,' is actually an abbreviation of the longer one *taueuta no sōshi*, 'bound book of songs for transplanting.' The expression *tauezōshi* is used in a number of ways by Japanese scholars. They can apply it to sequences of songs sung at the time of transplanting. Together these sequences comprise a type of *kayō*, performed poetry or song, described in the Preface, popular in the latter part of the Japanese middle ages and the early modern period. In English we might simply call them sequences of (trans-) planting songs. Of a large number of such sequences, some recorded for the first time only in recent years, a certain number are especially close in form and content to the particular text translated into English. Sequences of planting songs of this sort might be called texts of the immediate or direct tradition of *Tauezōshi*. In English, then, we can reserve the term *Tauezōshi* itself for the text translated here into English. The term will not be used, as it is in Japanese, of transplanting sequences in general.

The text translated into English, what we are going to call the *Tauezōshi*, is preserved today in two direct tracings, that is a copy made by placing thin paper over an original manuscript. One copy is kept at the Japanese Language Study Room of the Department of Literature of Tokyo University. The other, which is in most respects the same, is kept in the Historiographical Institute of the University of Tokyo. In print the text of the former tracing is the basis of what appears in volume #44, *Chūsei-kinsei kayōshū*, of the standard series of Japanese literature published by the Iwanami press in 1959. The text is edited and annotated by Shida Engi. An earlier printed text, 1928, appeared in volume five of a famous series of *kayō* edited by Takano Tatsuyuki. In this earlier edition Takano made his own surmises where the hand of the manuscript was hard to read and, in a sense to be discussed later, set the whole in order. No one seems to know where the original from which the tracings and printed edition were made is kept now, or if, in fact, it exists any longer. For that reason the question of the transmission of the manuscript used by Takano, from which the two tracings mentioned above were supposedly made, remains an open one. I want to give a general account of the problem and then to add several comments about the set of poems called a first set of morning songs which, though missing from both tracings, is present in Takano's printed version.

In the commentary to his printed edition Takano had this to say about the manuscript he used: 'This bound book for transplanting is one which has been traditionally used at the time of transplanting at the village of Shinjō in the district Yamagata of Hiroshima Prefecture. I borrowed it from Yamamoto Nobuki for my collection.'

Sasaki Nobutsuna, an eminent scholar of Japanese literature, had some relevant remarks to make about the same manuscript, the one now translated into English, in his own work *Kayō no kenkyū* (1944): 'Thanks to the kindness of Yamamoto Nobuki I have been able to see a volume of transplanting songs preserved in Hiroshima.' He goes on to add: 'An old copy of this volume for transplanting is in the collection of Mukai Matsutarō of Ōasa, Eda no Miya. It is a book of forty pages on Japanese paper. The beginning and end are missing. Supplemented by additions made through the help of Takeeda Ōmachi it amounts to five hundred thirty-five poems. [Sasaki's method of counting differed from that used in the Iwanami series and in the English translation.] Many are still sung today at the villages of Shinjō and of Ōasa. Some forty-eight or so of these songs are known by everyone in these villages.'

A comparison of the two passages makes it clear that both Takano and Sasaki saw Mukai Matsutarō of Ōasa, Eda no Miya's bound book of transplanting songs which was then in the possession of Yamamoto Nobuki. This is the version which was published in volume five of Takano's great collection of *kayō*.

Another comment on the manuscript contemporary with the two above comes in an essay of 1930 contributed by Fujita Tokutarō to volume 2, number 1, of a periodical called *Minyō ongaku* (*Folk music*): 'Some two or three years ago I was shown an old volume called *Tauezōshi* by Dr. Sasaki Nobutsuna. I liked the old quality of the manuscript. It was said to be a volume in the collection of Yamamoto Nobuki. A traced copy of this volume came into the possession of the Illustration (*sic*) Study Room of the Tokyo Imperial University. Later the work was included in the fifth volume of Takano's collection of *kayō* so that now anyone can see it.'

Though the expression 'Illustration Study Room' is suspicious it seems certain from Fujita's account that the so-called *Tauezōshi* which Yamamoto had is the original both of the two traced copies now at Tokyo University and of the version published in Takano's collection of *kayō*. The date when one of the tracings was made is recorded at the end of the manuscript in the possession of the Historiographical Institute.

> *Tauezōshi*/Brought by Yamamoto Nobuki/Copied in December of 1923

This dating accords with the comments made by Sasaki and Fujita: the manuscript of *Tauezōshi* was brought to Tokyo some time before December 1923.

In the spring of 1965 and 1966 I visited Shinjō and had a chance to look over various books of transplanting songs and works commenting on the tradition gathered together by Hisaeda Hideo of Shinjō. Among these was a volume of transplanting songs called *taue utazōshi* from the library of Kubo Toyomatsu. According to what I learned from Mr. Hisaeda this volume was originally in the collection of Mukai Matsutarō of Ōasa. Kubo Toyomatsu (born in 1897), another person of the area, at seventeen began to learn how to perform the role of *sanbai*, the leader of the transplanting ritual. In order to do this he borrowed an old and worn set of transplanting songs from Mukai and began to make a copy of this for himself. (As I said, the copy he made is now in Mr. Hisaeda's collection.) No sooner had Kubo begun to make the copy for himself from the volume which he had borrowed than Mukai asked him to return it. Kubo's volume is dated the end of May 1912. Kubo said that Mukai asked to have it back in order to lend it to a certain Takeeda, then an official at the Yasukuni Shrine in Tokyo. Kubo's copy goes only as far as the end of song eighty-four as numbered in the Iwanami series. Eight blank pages follow. Evidently this was as far as he got in copying before he had to give the original back. I have not yet had a chance to verify whether the Takeeda mentioned by Kubo is the same person as the Takeeda Ōmachi written about by Sasaki.

紅葉二首

一 うら兆口乃 るまで戸とひを不小 や見たれは
こかねにましなる あさひさしも
はるのわさひさしてるよ けるわきひや
秋はさいてはひかし 雲にかくれて

二 けさうつのときに 雲のもをみな遠ぎ
きり雲ろみす 雲ろ山のなをみたれも
きりがふかうてみせんれうしをたちまふ
あさのそりいてろうかなためのいれうかー
わかがふろうよてみせんれいての秋くさい

This account makes it clear that the original *Tauezōshi* was first of all transmitted in the family of Mukai Matsutarō. It passed from Kubo's hands into those of Takeeda. It was in the hands of Yamamoto Nobuyuki then living in Tokyo, who was a native of the same region in which Shinjō is located, what was once called the older province of Aki. It was seen by two specialists in Japanese literature and *kayō*, Sasaki and Takano. It also attracted the attention of Fujita Tokutarō. What later happened to the original is unclear. This is the state of our knowledge of the manuscript tradition at present. The task before scholars now is to attempt to reconstruct the original upon the basis of the two nearly similar direct tracings (or what amounts to about the same thing, on the basis of the version in the Iwanami series) and a comparison with the version printed in Takano's collection and Kubo's volume.

Other texts play a role in studies of *Tauezōshi* and of planting song sequences in general. These are ones which record sequences very close to the form of *Tauezōshi*. In fact we can think of them together as comprising a single group which I usually refer to as *kayō* in the immediate or direct tradition of *Tauezōshi*. My definition is not meant, however, to minimize the importance of a somewhat different group called the Bingo branch of transplanting songs and named after an area where this second type is prominent.

THE VARIOUS texts which together comprise the immediate or direct tradition of *Tauezōshi* begin with a first set of morning songs and end with a special group to conclude the day's work, *agari uta*, after the fourth set of evening songs. Both the version in the possession of the Japanese Language Study Room and that in the Historiographical Institute of Tokyo University lack a first set of morning songs. The Kubo version and that in Takano's collection both contain a first set for the morning.

Version in Takano's collection of Japanese *kayō*

First set for the morning

To begin the song, first sing of the field god
At the beginning of the planting, offer it to Inatsuruhime

The field god, *yāre*,[1] where does he come from, *yara*, from the shrine
From the shrine, *yāre*, on a white horse with the reins in his hands
With the reins in his hands the field god is on his way
On a white horse, let's get the field god going

At Yoshino is a village. In the village a grove. The grove bends
Yes, the grove bends down. The grove bends with night dew
Dew is deep. The grove at Yoshino bends
Let a breeze blow. Let the grove at Yoshino bestride it

Tubs of rice, eight. Tubs and rice scoops nine
. . . . . Chiyo . . . . .
. . . . . [2]rice scoops piled up nine
. . . . .
. . . . .

With what water did you wash the dishes used this noon
Washing with water from the valley, drying with figured cloth, we put them away
Drying with figured cloth, we put them up on the sacred shelf

Kubo's version of *Taue utazōshi*

First set for the morning

For the beginning of our song, sing of the field god
Look at the scroll about the origin of the land
The tip of the halberd became dew, became the land. At the beginning of this country the god Izanagi made it

The field god comes, *yara*, parting the dew
With leggings of brocade, wearing a small field hat of figured cloth
Parting the dew, now the field god has come. In his small field hat of figured cloth now the field god has come

(*Yuri*)[3] Where is the field god coming from? From the shrine
On a white horse with the reins in his hands
Happy. Lucky. The field god has come. On a white horse now the field god has come.

(Between this last line and the beginning of the second set for the morning there is a blank space of about two lines.)

These two sets of poems, though each appears in its respective version as a first set for the morning, raise a number of problems. Scholars have already pointed out that both lack the special organization and unity we expect of a first set of morning songs. It is true that the first few songs of the Takano version are similar to what we find in other versions which comprise the immediate or direct tradition of *Tauezōshi*. The first two lines invoke the field god. Together with the following song about the white horse they are typical of a special type of song, sometimes referred to as *yakuuta*. This is a type of song prescribed by the nature of the day's ritual, here invocation. But the third song, 'At Yoshino is a village,' raises a problem. The song mentions dew. In that sense it is a morning song. But does it belong so early in the series? If we look at other texts of the immediate tradition we find it recorded, in one example, among the fourth set of morning songs, while in another it comes within a special group called 'the beginning of dew.' In another relevant work, one which records brief instructions on the proper order of items in these song sequences, it has a place in the first set of morning songs but much later in these. According to the instructions it should follow another group of songs, the one about adorning oneself, applying make-up, beside a thin paper partition which is represented in the English translation (stanza 17, for example). In another related version the song 'At Yoshino is a village' comes in the second set of morning songs. Though only a few preserved sequences include this particular song at all, when they are compared in this way they raise doubt as to the appropriateness of the song coming directly after the ritual verse about the white horse. It seems to belong, rather, to a group on the famous place Yoshino.

The next song, 'Tubs of rice, eight. . . ,' is the same type as number twenty-nine of the fourth set of morning songs given in the English translation, i.e. that of the Iwanami series. Most texts of these sequences place a song of this sort just before the series of noon songs, that is after the fourth set of morning songs. The content of the song, in fact, makes it clear that it should be sung just before the *unari* appear with food for the noon meal. It does not belong, as given in Takano's text, in the first set of morning songs. The last song, 'With what water did you wash. . . ,' is of a similar type. It is a song for washing the utensils used at the noon meal. One relevant text places this song in the first set of noon songs. In a famous modern anthology of folk songs called *Riyōshū* the song appears among a section of songs from Asa, a district of Hiroshima Prefecture, recorded as coming in the afternoon.

In short, what purports to be the first set for the morning in the Takano text is not appropriate for that time of day. As a matter of fact the very title itself, where the genitive particle *no* is inserted between the words for morning and for song, is questionable in the light of normal practice in *Tauezōshi*. Further, the expression noted in the margin at one place, 'just as it is in the volume,' makes it appear that Takano had seen a *Tauezōshi* version at a stage where the first set of morning songs had been added from one or another volumes of transplanting songs. Whatever particular volume(s) of transplanting songs may have been used to provide Takano's first set of morning songs, it is very likely that the Takeeda mentioned before had some part in the undertaking.

When we turn to the Kubo text we find that with the exception of the song about the god of the field there are serious problems here too with what claims to be a first set of morning songs. The verse 'Look at the scroll about the origin of the land' is based on a passage in myth about the birth of Japan. A song like this is more likely to be found in the Bingo group of transplanting songs, already mentioned as a collateral though distinct type. Such a song is not widely represented in the more narrowly defined direct tradition of *Tauezōshi*. For example, a representative text of the Bingo group has among its morning songs the following lines similar to those in the Kubo text:

> What god made this island? It is the island made by Izanagi

> Anciently he bore all the sea and land. . . . . . dew from the halberd of Izanagi
>
> Dew from the halberd, the dew collected and became an island. He disclosed the island Onogoro

When we consider the usual structure of the narrowly defined direct tradition of *Tauezōshi* it is suspicious to find lines like these coming before the famous verse about the white horse.

An investigation like the one we have been conducting shows that the version in Takano's collection and that in Kubo's volume have both added a so-called first set of morning songs which was missing at some stage from the original, as can be verified by the two direct tracings. In order to remedy the deficiency these two versions drew upon other books of transplanting songs. Further, it is clear that whatever books of transplanting songs were used for this purpose they did not follow a form close to that of *Tauezōshi* itself. We can imagine, then, that at some time after the original *Tauezōshi* had been in the hands of Mukai Matsutarō sections, including at least the first set of morning songs, fell out. Kubo for his version supplemented in his own unique way the text he was copying with a first set of morning songs drawn from other books of transplanting songs, as did Takeeda in his own way or Yamamoto or Takano himself for the other version.

Still it *is* possible finally to add here a version of the first set of morning songs which is an authentic one to judge from a comparison of books of songs for transplanting which are closest to what we have been calling *Tauezōshi*, that is texts in the direct tradition of *Tauezōshi* narrowly defined. A version which is found in a few song books, including a text called the *Takamatsuya kohon*, has already been judged sound in this respect by a number of scholars. As the following version, the *Takamatsuya kohon*, shows, a reliable version has songs from a group invoking the god of the field (songs of a type called *sanbai oroshi*), and ones from another group about a man's visit to the girl he wants to make love to (*koi no shinobi uta*).

Let us first offer rice seedlings to the god of the field

Let us offer the transplanting to Inazuruhime

Now *sanbai* comes. Down from the shrine
On a white horse with the reins in his hands
With the reins in his hands. Now he is up and riding
On a white horse, let us get *sanbai* going

Alas the night I waited but the one I love never came
I composed a poem. I composed a song. It brought no sleep
Did you hear something when you got up? A sound this morning like someone striking a drum
I'll make a song waiting for him to come

Alas the night with no kiss from him
I love him more and more, more and more
This morning, too, awakening he asked me for water for his hands
Water for his hands. Pure water where the rivers meet. Seven pails full

Oh the night he came and slept
Because we talked of love we had no rest
How angry it makes me. The dawn crow calls early
We say sleep. But that means only until the sun is beside our pillow
The one I love forgot and left his short sword behind in our room

We can see that the traditional form for a first set of morning songs is one, first of all, invoking the field god and then of songs about a young man's visit to make love (*koi no shinobi uta*). These two groups of songs are what

ought to be sung in a first set of morning songs. Songs of the type *sanbai oroshi* are sacred songs (*kamiuta*) invoking (literally 'going out to meet') and praising the field god. The later group, *koi no shinobi uta*, sing of the sexual passion of the young. Implied in this, or behind it, is the idea that fulfillment of sexual desire through the physical union of man and woman influences, sets in motion, the growth and the development of the rice crop. Songs for invoking the field god and songs of sexual fulfillment are at the same time sacred songs and songs of what we might term magical power. (This latter type of song is called in Japanese *juka*.) The first set of songs for the morning is composed through sacred songs and *juka*. The structure of the next set, the second set of morning songs, is an extension of these aspects of the first.

[1] *Yāre, yāra.* These are words without meaning, sometimes called *hayashi kotoba*. I have not included them in my translation of the main text though they are sung in performance.

[2] Here, as in the main text, a ". . . . ." marks places where there is not yet a generally accepted interpretation by Japanese scholars.

[3] (*Yuri*). One of a number of indications of the manner in which the songs were sung which I have omitted from the translation elsewhere.